ANIMAL GENIUSES

CROWS

THERESA EMMINIZER

PowerKiDS press

Published in 2026 by The Rosen Publishing Group, Inc.
2544 Clinton Street, Buffalo, NY 14224

First Edition

Editor: Theresa Emminizer
Book Design: Michael Flynn

Photo Credits: Cover, p. 1 Shaynak Photography/Shutterstock.com; (series background) Viktoriia LiSa/Shutterstock.com; p. 5 Cristian Gusa/Shutterstock.com; p. 7 Karen Hogan/Shutterstock.com; p. 9 Maksim Ladouski/Shutterstock.com; p. 11 JakubD/Shutterstock.com; p. 13 Brian E Kushner/Shutterstock.com; p. 15 Brad Balfour/Shutterstock.com; p. 17 colourbird/Shutterstock.com; p. 19 Hyung min Choi/Shutterstock.com; p. 21 kiran aadil/Shutterstock.com.

Cataloging-in-Publication Data

Names: Emminizer, Theresa.
Title: Crows / Theresa Emminizer.
Description: Buffalo, NY : PowerKids Press, 2026. | Series: Animal geniuses | Includes glossary and index.
Identifiers: ISBN 9781499451825 (pbk.) | ISBN 9781499451832 (library bound) | ISBN 9781499451849 (ebook)
Subjects: LCSH: Crows–Juvenile literature. | Crows–Behavior–Juvenile literature.| Animal intelligence–Juvenile literature. | Crows–Psychology–Juvenile literature.
Classification: LCC QL696.P2367 E45 2026 | DDC 598.8'64–dc23

Manufactured in China

CONTENTS

Clever Corvids

American crows are big black birds that are common throughout the United States. They're part of a group of birds called corvids. Crows are well known for the loud "cawing" sound they make. They're also well known for being clever, or smart. In fact, crows are some Earth's smartest animals!

A Murder of Crows

Crows have close families of up to 15 members. Young crows help their parents raise new baby brother and sister crows! They help their parents build nests and bring food to their mother while she sits on new eggs. A flock, or group of crows, is called a murder!

Crow Funerals

When a living crow finds a dead crow, it holds a crow "**funeral**!" The living crow calls loudly to other crows nearby. The crows gather around the dead body, making a lot of noise! **Scientists** believe crows do this to learn about what threats, or dangers, are nearby.

Crafty Crows

In storybooks, crows are often pictured as tricky or crafty. That's true! Crows may steal food from other animals or from people eating outside. They work together as a team to catch an animal's eye while taking their food! Crows eat almost anything, including trash and dead animals.

Using Tools

Crows find clever ways to make and use tools. One crow was seen using a cup to gather up water and carry it over to their dry food to make it wet! Crows also use sticks to spear, or stab, bugs or get things that are hard to reach.

Solving Problems

Crows find clever ways to solve, or find answers to, problems. They are even able to plan ahead several steps! Crows drop hard-to-crack nuts on busy roads. The cars run over the nuts, cracking them open. Then the crows fly in to gather their food!

Remembering Faces

Crows **recognize** faces! Crow communities have long **memories** and let one another know about people they like or dislike. In one study, masked **researchers** caught seven crows in nets. They let the crows go. But for the next 10 years, crows in that place would caw and dive at people wearing those masks!

Trouble with Crows

Crows roost, or settle to rest, in very large groups. There can be hundreds of birds in a roost! Roosting may cause trouble between crows and people. People also clash with crows over crops, which crows try to eat. There are many ways people can keep crows away without harming them, such as putting up scarecrows!

Be Kind to Crows

Because crows are smart and playful, some people keep them as pets. Like parrots, crows can be taught to make noises that mimic, or sound like, words! But crows are happiest in the wild with their families. You can befriend a crow by quietly watching them or leaving them snacks.

GLOSSARY

funeral: A ceremony or event that's held because someone has died.

memories: Thoughts of past people, places, or events.

recognize: To see and know someone or something.

researcher: Someone who does research, or careful study to find new knowledge.

scientist: A person trained to study the world around us and make discoveries.

FOR MORE INFORMATION

BOOKS

Barnes, Rachael. *Crows*. Minneapolis, MN: Bellwether Media, 2023.

Pang, Ursula. *Crows*. Buffalo, NY: PowerKids Press, 2025.

WEBSITES

Britannica Kids: Crow
kids.britannica.com/kids/article/crow/353020
Learn more about the amazing crow!

National Geographic Kids: American Crow
kids.nationalgeographic.com/animals/birds/facts/american-crow
Find out more fun facts about crows!

Publisher's note to educators and parents: Our editors have carefully reviewed these websites to ensure that they are suitable for students. Many websites change frequently, however, and we cannot guarantee that a site's future contents will continue to meet our high standards of quality and educational value. Be advised that students should be closely supervised whenever they access the internet.

INDEX